Introduction

There is nothing very strange about the scene above – it could be almost anywhere in our countryside.

It is easy enough to take it for granted. Horses drinking from a pond. It is as simple as that.

Or is it?

When you begin to ask questions, things often turn out to be not quite as straightforward as they seemed at first. As a start, why is that pond there? You know what happens when you pour a bucket of water on to the ground. It soaks away and disappears. What makes a pond permanent?

Look at the plants flowering in the pond. They are colourful – and they attract many flying insects to pollinate them. *Water* plants pollinated by *air-borne* insects . . . there is something puzzling about that, if you think about it. . . .

So, in this **WATCH Book** we are going to investigate water in a fresh way. You do not need to be an expert naturalist – there are plenty of easily recognised signpost plants and indicator animals to help you decipher the story.

And there is plenty of water to look at! Ponds, lakes, rivers, streams, marshes, boggy places, and even those puddles that dry up in a few days – water is everywhere. Indeed, there is an old saying that water is 'the eye of the countryside'. That sums it up in a way.

So let us find out what it is all about.

Ancient water

How old are the ponds and lakes, rivers and streams? Are they as old as the hills? Are they as old as the Romans?

There are some clues in the picture *above*. It is a hamlet in the Lake District – a part of Britain that was being scraped and sculptured by glaciers of ice during the Ice Age, which ended 10,000 years ago.

The tough rock of that Lake District hillside has been smoothed by ice. Since then, rainwater has collected into a stream which has eroded the gully you can see.

The unlocking of the ice meant that vast amounts of water were released. The sea level rose, and eventually the English Channel flooded, and the British Isles took on the shape they are today.

Inland, ice sheets had been smoothing the lowlands too, in some places, pushing debris from the hills across them. But as the icy grip eased, our main rivers established themselves more or less where you can see them today.

The Lake District view opposite *contains clues to the way our landscape has been shaped – by ice and by water. Notice the water-worn gully in the smooth ice-shaped flank of the valley.* Below, *a river at work today.*

Look for evidence of water erosion, still taking place. Look at the bend of a river or stream. On the outside of the bend, the bank is being cut away. On the inside, debris is being deposited, sometimes making a beach as you can see *below right.*

Moving water has energy – to erode and carry off the erosion debris. How much energy it has depends not only on how much water there is (the size of the river) but also on how fast it is moving. Water moving *twice* as fast can move rocks *thirty* times bigger. That has an interesting implication:

Look at the banks of a river or stream. Why are there banks at all? When are they cut – by the placid *summer* or the rushing *winter* water?

How fast is the water moving? Drop in a piece of wood and time it over 100 ft (about 30 m). A rate of 1 ft per second (about 30 cm per second) will carry fine silt – the particles will not fall straight to the bottom if you drop some in. At 4 ft per second (about 1·25 m per second) the water can disturb rocks weighing 20 lb (about 10 kg).

So, when a stream is running down a hillside, it is a force to be reckoned with. Often, however, the rock of a hill is in layers, some harder than others.

Look at a waterfall. Is there evidence that it forms where the water is eroding *harder* rock? Why is a waterfall often at the head of a steep sided gorge?

Water also dissolves minerals from the rock it is in contact with.

Find two small pieces of chalk rock or limestone, drop one into a glass of water, the other into a glass of fizzy soda water, and leave overnight.

Some limestones are tough and hard, but often you will find that the second piece is mushy outside; it has started to dissolve.

The fizz in soda water is carbon dioxide – a gas found in air. Rain absorbs it on the way down, making a weak acid, which dissolves the calcium carbonate of the rock.

Animals in the water can make use of that dissolved mineral. Water snails make their shells from it.

We also know it is there – we call the water 'hard' if it is.

Water tame

Running water has a tremendous force. It has helped to shape our scenery. Even valleys which are now dry (and there are many of them in the chalk downlands of the South) have usually been cut by water.

Ponds are placid. Indeed, that village pond *above* does not even look natural. How old is it? Is it man-made?

It is a fact that most of the ponds we see *have* been made by man. Sometimes all that was needed was to enlarge a smaller, natural pool taking advantage of the watertight clay layers already under the soil. Sometimes a stream was dammed. Often, however, the pond was made by digging a hole and lining it with clay; it then filled with rainwater.

Villages were founded near a reliable water supply – around an enlarged pool, maybe; or by a stream. But water was also needed out in the fields for the cattle, and ponds had to be made. A hundred years ago there was a field pond shared between every two or three pasture fields. Today water is taken to the cattle by pipes, and the ponds have often disappeared.

The village pond opposite *looks much too tame to be natural – is it?* Below right, *a waterscape that is man-made, though it does not look like it: the* **Norfolk Broads** – *the Broads occupy old peat diggings.* Bottom, *a 200-year-old* **canal**.

Look for ponds out in the fields. You will not find many left, but maybe you can find clues – a hollow in the ground, perhaps marked by bushes that once grew at the pond's edge.

Look for specially built WORKING PONDS – mill ponds, hammer ponds (which provided water for forge machinery). Look for pits dug for marl, now flooded.

Some open water created by man now seems natural. The Norfolk Broads are man-made. Peat was a useful fuel in medieval times, and thick beds of peat were dug in Norfolk. The diggings then flooded – to form the Broads we know today.

Some canals are now a couple of hundred years old. Like many ponds, they were dug and lined with clay. If you can find stretches of canal on the map, go and visit them. They are now often partly wild, as the view from under the bridge (*left*) shows.

However, canal boating is a very popular holiday, and when canals are still being used, they are usually kept cleared.

Look on the map for other watery clues, in village names. 'Lower Swell' is a village in Oxfordshire, you can guess how the name arose.

'Bourne' is sometimes part of a village name. It means stream (the word *burn* is similar). In Wiltshire, four or five villages are called 'Winterbourne'. The local rock is porous chalk; the streams only flow in winter, when the chalk is completely saturated with rain.

'Brook', 'ford', 'marsh' – there are many watery elements in village names!

Water life

When you walk up to a pond or stream, you will often hear a sudden 'plop'; you've disturbed a water vole *above*. If you move fast, you may get a glimpse of it swimming underwater, shiny with a coating of small bubbles over its fur. But it is quick, and usually it will be gone, hidden by the plant stems or gone into its burrow in the bank.

The water vole often leaves a pile of chopped stems and leaves on the bank, ready to eat. Another clue is a patch of closely cropped 'lawn' on the bank.

The water vole is a harmless animal. It is often called the water rat, but it is not a true rat – it has the vole's snub nose, stubby ears and it has a short tail.

It is so common that we take it for granted, yet it would only be there if certain conditions were met – undisturbed bank to burrow, plants for food. Yes, plants come first in the natural world. So let us start with them.

The **water vole** opposite *(also known as the 'water rat') is a common inhabitant at the banks of rivers and ponds. The striking looking* **arrowhead**, below right *(seen here ⅓ natural size) is a good plant to start our investigations with.*

Plants come first

Plants come first in nature – without them there could be no animal life.

Plants lock up the energy of sunlight, using it to manufacture sugars in the green leaves. Water and the gas carbon dioxide are the raw materials, and oxygen is released. Animals reverse the process – they burn sugar with oxygen to provide energy, and they release carbon dioxide.

Plants are well engineered. Land plants have a strong stem or trunk to carry the leaves; and they have roots for anchorage and to collect water from the soil.

Are water plants different?

Look at the arrowhead, *right.* It is an exciting looking plant, found in slow moving or still water. It is rooted in the mud at the bottom, but its stems are strong and rise above the surface.

But look at the leaves. There are:

- long grass-like leaves underwater.
- blunt oval leaves that float.
- arrowhead leaves carried only on the aerial stalks, which sometimes rise a metre above the surface.

Why are the leaves different? Are the grass shapes better able to survive the tug of the current? Would flat leaves tear in the current?

So, a water plant may have to cope with two sets of conditions at least. That may be obvious, but. . . .

Look at the *flowers* of the arrowhead. Where are they? Like land plants, they are pollinated by airborne insects.

That fact has a rather surprising implication.

The different-shaped leaves of the arrowhead are a good example of the way that plants *are exactly engineered* to suit their setting. So does the fact that the flowers are pollinated *above* the water mean that the arrowhead is not really a true water plant?

Pond plants

Many people look on water as a 'simple' environment, from which life on planet Earth originated and evolved. At some late stage, living things, plants and animals, left the water and colonised the land.

Well, that may be true as a picture of far distant times in the past. But does it have to mean that the same is true today? Is water life 'primitive'?

There is plenty of evidence pointing to the opposite! If we start with water plants we find that many are flowering plants, pollinated in air. It seems that they have colonised the water *from* the land! How they have adapted themselves to the new environment is fascinating; but the fact remains that they flower in air as a reminder of their origin.

However, there are some differences. They establish from seed less easily, often having to rely on underground stems to spread themselves. Certain land plants (nettles and blackberries are examples) also rely on spreading themselves underground in addition to seeds. *Annual* plants, which are common on land, are rarely found growing in water. It is not difficult to deduce why.

So let us start with the picture *above*, of small duckweed leaves and larger water lily leaves floating on the surface of still water.

Each duckweed 'leaf' is a plant; if you picked one up, you'd see small roots dangling down underneath. Duckweed can carry minute flowers, but buds also grow from the side of each plant; they separate off, and grow into new plants. Duckweed can cover a pond without a single flower.

Opposite, **duckweed** *and larger* **water-lily** *leaves on the surface of a pond. There are three or more different duckweeds, distinguished by the shapes of their leaves.* Below, *a sketch showing different kinds of plants to look for; and* **yellow water lily**.

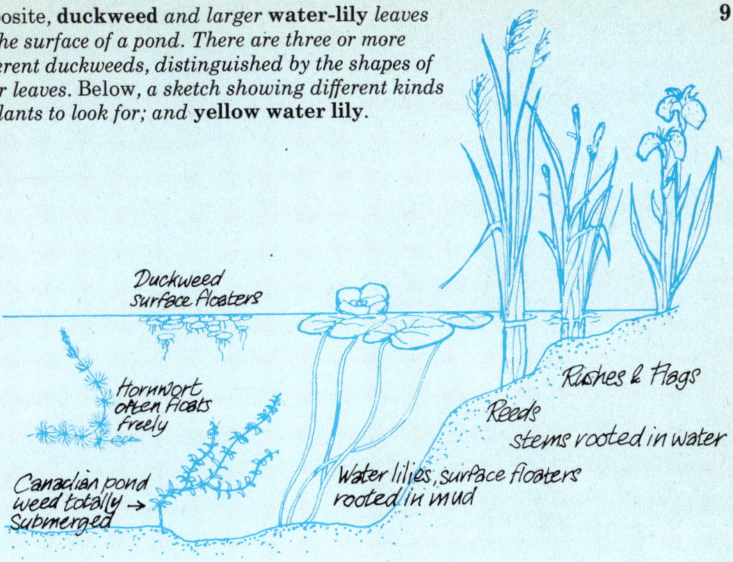

If you feel down below the surface, you'll find that the water lily is rooted in the mud at the bottom. The leaves grow on very long leaf stalks.

And you will find other plants in the pond, growing totally submerged. They do not reach to the surface.

So there are *three* different kinds of plants growing in open still water:

- free floaters, like duckweed.
- surface floaters rooted to the bottom, like water lilies.
- totally submerged plants.

Pull up one of those totally submerged plants. What kind of leaves do they have? Can you find any evidence that they flower?

The life of those plants is a good fit to their environment. Winter ice is a threat. Duckweed leaves store starch in summer, and by autumn they are heavy, and sink. The starch reserves are used up during winter, making the leaves lighter; they float again in spring.

Water lilies have a different strategy. In winter they die down to the rootstock in the mud at the bottom of the pond. (Many land plants also die back in winter.)

Water crowfoot, top *(the photo is ½× life size) is one of the buttercup family. At* bottom, **water bistort** *with pink cylindrical flower heads standing clear of the water surface. The flowers are seen in late summer.*

Look for water crowfoot, *above*. It is a small, white buttercup, with aerial flowers often massed like a snowdrift on shallow ponds.

Crowfoot grows from roots in the mud of the pond. Its surface leaves are flat and lobed, as you can see, but those underwater are like threads.

Crowfoot can survive drought – it grows in a dwarf state if the pond dries up.

Can any other 'water' plants survive out of water?

Look for the water bistort, *below*; it is a typical water plant, with aerial flowers and oval surface leaves as you can see. It has *very long* stalks reaching down to its roots at the bottom – sometimes 2 m long.

However, some botanists claim that they have found a form of water bistort growing away from water – with silky, willow-like leaves and few flowers.

That is another clue that the pond's surface may not be the clear-cut boundary we assume it to be.

The sketch below *shows the characteristic profile of* **hornwort**; *an exception because it does flower underwater.* Below right, *the* **yellow flag** *(wild iris)* $\frac{1}{2}\times$ *life size; it is a signpost to marsh conditions at the edge of the pond.*

Here is what you may find:

REEDS – *tall grasses*, with flat leaves and grass-like flowers, often growing in dense beds in open water. Reeds can be seen in the Norfolk Broads picture on page 5.

RUSHES – Tufted plants with long narrow cylindrical leaves. The flowers are not showy, and the flower stem is usually round and *pithy*.

SEDGES – grass-like, but the stem has a *triangular*, three-sided cross-section (grasses have a round cross-section).

Look also for other plants, maybe 'flags' (wild iris), *below*.

Underwater leaves, whether they are on plants rooted in the bottom or on plants floating freely beneath the surface, are usually thin and threadlike, or split up into a bundle of threads.

The fact that all plants show the same trend is a good example of 'convergent evolution' – the thread is a *successful* leaf shape underwater.

Look for hornwort. Its distinctive clusters of leaves are sketched in the picture *above*. It is an unusual plant, because it is one of the few flowering plants that does flower *underwater*. The flowers are very small, and the pollen is carried from one to another by water currents.

So much for the plants and flowers of open water. Let us now look at the plants growing nearer the bank. There could be quite a jungle, at the edge of an undisturbed pond or shallow lake.

Swamps and marshes . . . and dry land

Open, fairly shallow and still water will always become dry land, if left alone for long enough. Indeed, there is a clue in the picture that began this investigation! Have a look at the pond in the picture on page 1. The tussocks of sedges *above* are another clue.

Reeds can grow out in quite deep water, but their stems trap silt, and their dead leaves and stems sink down – so that in time a bed of peaty litter builds up. The water becomes shallower, and other plants (*sedges* and *flags* for example) can root in the soggy mush, and compete with the reeds. The litter continues to accumulate until it is at the water level for most of the year – except during winter floods maybe. *Tussocks* of sedges can then grow, and in time the seeds of woody, wet-loving shrubs can grow, and scrub-covered dry land is the end result. The events are sketched *opposite*.

Much the same kind of thing can occur at the edge of any undisturbed pond.

Opposite, **sedge tussocks** *in winter. They signpost an advanced stage in the story of open water.* Below, *the well-known 'bulrush' and water mint; and the sketch shows how plants identify the conditions you can find.*

Look for the 'bulrush', *left*, a signpost of *swamps*. Its real name is the great reedmace. However, ever since a Victorian painter used it for a picture of Moses among the Rushes, it has been called a bulrush! Odd botany!

Look for the water mint, *below*, and other lushly leaved (and often smelly) plants of the *marsh* stage of the story.

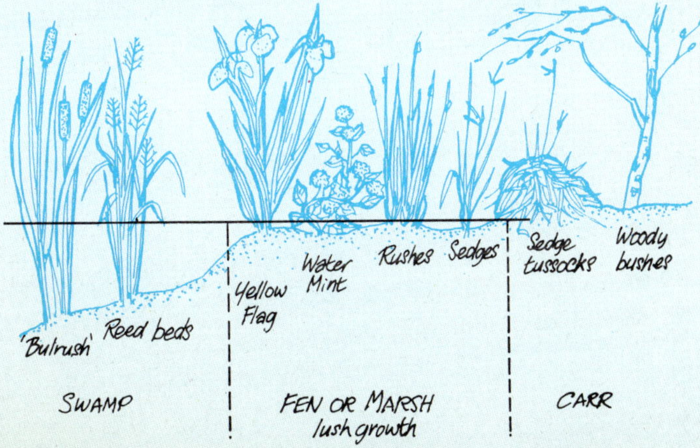

Bogs

The words 'swamp' and 'marsh' are so familiar, that perhaps it was a surprise to find that they had a precise botanical meaning. Is it the same with the word 'bog'? What does it mean?

A *bog* is, like the other two, a name for a *community* of plants; a collection of sometimes many different kinds of plants growing at one place. *Heath*, *wood*, *grassland* are words for yet other plant communities.

Bogs cannot be found everywhere, but only where it is very wet: on rain-sodden hillsides, or in waterlogged hollows in valleys and lowlands. Nowadays, extensive bogs are found only in remote places. In lowland Britain most of them have been drained, but small patches can still be found scattered about.

The dominant plant of a bog is bog moss. It can survive being waterlogged with a poor, acid, mineral supply (indeed, it will die in chalky water).

Extensive bogs, which can still be found in Scotland, Ireland, Wales and northern England go through a 'life cycle'. The bog moss flourishes and builds up thick layers of peat with its dead stems. In time hummocks develop, which are too dry for the moss, but in which cotton grass (see the picture *above*) or heather can take root. The dry hummock stops growing but is eventually overtopped by neighbouring hummocks, becomes a hollow, fills with water – and the moss can grow again.

Cotton grass *is growing on the bog* opposite, *a sign that it is drying up*. Below, **bog moss** *(in close up), the dominant plant of bogs, and* at bottom **sundew** *(life size), which obtains its mineral supply from insects trapped on its sticky leaves.*

Look for boggy patches and inspect them. You may find them at the edge of a pond, or in a ditch, but the best place to look is on moorland.

Look for bog moss (*right*). It is not a flowering plant. The stems are very weak, so that each is supported by its neighbours.

Look for sundew, *below*, growing in bogs. It is unmistakable. Its reddish leaves are held in rosettes. Each leaf is studded with hairs which glisten in the sun. The sparkle attracts small insects which stick to the hairs; their bodies are then dissolved in enzymes exuded by the plant and digested.

In that way, sundew can obtain minerals which it cannot gain from the very 'soft' water of the bog.

The sundew is a flowering plant. The white flowers are on a central stem, springing from the centre of the leaf rosette.

Other flowering plants are also associated with bogs:

Look for bog asphodel – its conspicuous spike of *yellow* flowers is most beautiful – a good signpost to boggy land.

Vast bogs are nowadays only to be found in a few places. In others they did exist – but have long ago been dug for the peat under them, drained and turned into farmland. Or building land!

Look on the map for names which recall the old peaty bogland. 'Moss' is a common name in the North, and around Manchester.

Closed worlds? Part 1

When we stand on the bank of a pond or stream looking in, we see what we take to be a completely separate world. The environment is very different from that of land. The water surface is a barrier between the two worlds, marking the difference.

Yet that surface is not, of course, anything like an impenetrable barrier. Many animals cross it freely.

So, could it be that the surface is not quite the definite division we assume it to be? We have already discovered some unexpected things about water plants – in the distant past, many of them seem to have left the land to colonise water. What is the score with the animals?

As an echo of their origins far back in the evolutionary story, many insects spend their younger, formative stages underwater, emerging into the air only when they are mature and able to mate. They then lay their eggs in water.

The dragonfly *above* has been snapped just when she is scattering her eggs in the pool by skimming over the surface. Dragonflies (and their relatives the damselflies) mate head-to-tail, flying in tandem. Sometimes the eggs are laid while the pair is still coupled.

Two of the many **caddis-fly larvae** *to be found are shown on this page. They are both about 3× life size. Not much is known of their way of life. The adult caddis is rather like a clothes moth (but it is not a moth).*

How to dip

Mayfly can only live in very clean water and give us a way of telling if a stream is clean – and likely to contain trout!

Mayfly and many other small animals can be found quite easily, by dipping. All you need is a home-made net, made by running some stiff wire around nightdress or curtain material (with about 1 mm between the strands), and tying it to a stick. You need a dish.

Put some white paper in the bottom of the dish and fill it with clear water. Then skim the net over the mud and through the water plants under the bank, and lift it out. Let it drain for a few seconds, then empty its contents into the water in the dish. In a few minutes what you have caught will have recovered and begin to move.

Look for the mayfly nymphs – the clean water indicators. Do not confuse them with the three-tailed damselfly nymphs. The mayfly nymphs are usually smaller, and if they swim, they move their bodies *up and down* through the water, *not* from side to side in fish fashion. However, many mayfly nymphs will be found crawling.

Look for stonefly nymphs – they are rather similar to mayfly nymphs, but have two tails. They crawl very slowly, and are sensitive to pollution.

Look also for many other animals, for easily recognisable *freshwater shrimps*, for *small red worms*, and for the *water louse* – looking rather like a woodlouse. They should all be very numerous, but as they can stand some pollution they may be found when mayflies and stoneflies are absent.

Look for a caddis-fly larva. They build tubes from many things: sand, pieces of leaf, even shells as you can see *above* and *below*. Not very much is known about their habits.

Look also for animals shown on the next few pages of this **WATCH Book**! If you have not got a net, many of the animals can be found living under stones on the stream bed.

Closed worlds? Part 2

Some animals eat plants. Other animals eat those first animals (or each other). Some animals live by scavenging dead plants and animals.

But plants come first. There cannot be any animal life, small or large, without plant life of some kind. There cannot be more feeders than there is food to feed them! That is true of land and water.

But there are subtle complications in a pond which we do not meet on land. Animals need oxygen which plants produce when they synthesise sugar. (Oxygen can also enter from the air directly through the water surface.) When animals breathe they use that oxygen and release carbon dioxide. The amount of oxygen water can hold also depends on its temperature.

As a result, there is a complex daily pulse in a pond, when the chemistry of the water changes, and in a small pond it can be a strong pulse.

In other words, although the surface is not quite the rigid boundary many people think it is, it does mark off a world with rules of its own.

You can also find water animals which feed in a way unknown (and impossible) on land. The swan mussel *above* is a *filter feeder*. It lies half buried in mud, pumping maybe 20 gallons (80 litres) of water an hour over its gills, which rake out food particles.

That mollusc could never have lived on land! Yet many of its relations seem to be recolonising water, from a terrestrial existence.

The **swan mussel** opposite *is a natural vacuum cleaner (and is useful in keeping the water clean in a garden pond).* Below, *a* **wandering snail,** *found in all kinds of watery places. At* bottom, *a slime of* **eggs of water snails** *(1½ × life size).*

Look for water snails. They are molluscs, related to the mussel. They are commonest in hard-water areas – they use calcium carbonate in the water to build their shells. They feed by grazing green slime from the surface of plants or rocks. They glide along on a trail of slime.

You will often see the snail *above* extend its trail under the surface, and move along upside down, as if across a ceiling.

It goes to the surface to obtain oxygen – to fill a *lung*. Yes, it is an air breather, like its cousin the garden snail.

If you disturb one of those snails at the surface, it will sink like a stone. It can survive (for a short time in the active warmer months, but for longer in the dormant winter) by absorbing, through its skin, oxygen dissolved in the pond water.

Other snails are better suited to life underwater; they have a *gill* which efficiently extracts oxygen dissolved in the water. They are easy enough to recognise – they can pull themselves back into the shell and close the opening with a round plate, rather like a winkle.

What are we to make of the water snails? If life did originate in the sea and then colonise the land, are the snails recolonising the ponds and rivers? The lung snails obviously are. But what about the snails with gills? Is the gill the ancestral method of obtaining oxygen, or has it evolved from the lungs of the other snails?

Another look at the surface

Water has a skin where it meets the air – and that is important in several ways.

Not to the water vole, perhaps, but to many small animals. We have already seen that some snails can treat the surface skin like a hard ceiling! Other animals use it in other ways.

Pond skaters are familiar animals – they are not floating; they support themselves with their fine legs on the surface skin. Look too for dusty-looking but energetic gatherings near duckweed – the 'dust' is swarms of minute 'springtails' which can also walk on water.

And look for the unmistakeable whirligig beetles *above*. They have shiny black wing covers. They weave around each other at incredible speed (faster even than a spider scuttles). They are hunting.

They locate their small prey (and avoid each other) by interpreting the ripples on the surface – they move with their antennae touching it. But they can also dive after prey – and they have bifocal eyes, one half suited for seeing in air, one half for seeing under the surface.

Whirligig beetles opposite *are common on many ponds weaving around at high speed. The* **water boatman** bottom below *is a predator. At* top below, *the* **china mark moth** *– its markings like potters' marks. Its caterpillar lives underwater.*

Look for a china mark moth, *right*, at rest on a waterside plant. It is interesting.

Butterflies and moths are completely terrestrial – unlike other insects there is no echo of a watery ancestry in their way of life. But the caterpillar of the china mark moth has recolonised water – it lives under the surface.

The eggs are laid on water plants (often water lilies), and the small caterpillar at first burrows into the leaf. Later, however, it cuts pieces about 3 cm long, and binds them together with silk, to make itself an oval case fastened to the underside of the floating leaf. The caterpillar lives in the case, sticking its head out to feed.

It keeps an air supply in its case. And the water surface has been crossed once again! Some species of china mark moth also *pupate* (go through the chrysalis stage) underwater.

Look for the water boatman, *left*. It is a *bug* – an insect with sucking (rather than biting) mouthparts.

The water boatman has a nymph stage which looks very similar to the adult.

The boatman swims upside down. At the tail end of its body are two pores through which it takes in oxygen – from the air. So you will often see it, upside down, its end seemingly stuck to the surface of the water. It is a typical posture.

The boatman is a fierce predator, hunting other insects or tadpoles, or even small fish, and it can swim very fast, rowing itself through the water with the oar-like legs that you can see in the picture.

It has very fast reactions. It needs to have, because you will often see it row itself quickly up to the surface, break through it, fly, and then dive back underwater again! It folds and unfolds its wings in a flash, and exactly at the right moment.

The **great diving beetle** top left, *life size, is a female; the male has a smooth black back.* Below *her is the* larva *(3× life size) – a predator that will attack tadpoles and fish.* Right, *a* **water scorpion** – *a predatory bug with sucking mouthparts.*

Look for a diving beetle, *above*, and its larva, *below*. Both are fierce predators – and both come to the surface for air, sticking their back end up through it.

Diving beetles fly from pond to pond, usually at night – and presumably navigating by the light reflected from the surface of the water.

Look for a water scorpion, *above*, hanging from the surface, breathing like a diver through a tube. It is a fierce predator, its front legs shoot out to catch its prey.

It is much less common than diving beetles, although it is just as successful a hunter. So the reason may be that it cannot cross the surface to fly to other water when the pond it is in dries up!

The **water mite**, *seen* top right, *8× life size, is related to spiders.* Below, *a* **raft spider**; *this one is relatively common, but one species is so rare that a patch of relic fen has been made a special nature reserve to save it.*

Look for water mites. One is shown, much magnified, in the picture on the *right.* Some mites are bright red – you can see them moving as active dots in ponds in summer. And you may, if you have good eyesight, see them infesting a pond skater!

Mites are related to spiders.

Look for spiders in or on water. The *water spider* spins a hollow bell for itself, which is anchored to leaves below the surface. The spider fills it with air by bringing down bubbles from the surface (and the air gives the bell a shiny look). It swims with a clumsy paddle.

Another kind of water spider is the *raft spider* seen *below*; it is not common, but can be found in the New Forest. The raft spider is one of our largest spiders. It waits on floating leaves, and darts off after its prey, walking on the surface of the water like a pond skater.

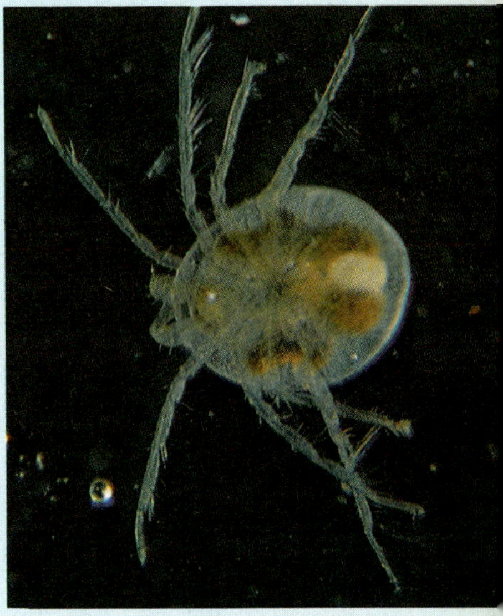

So, just as with the plants, there are quite a few examples of animals that have returned to the water, colonising it from the land. Some seem to be very successful in their new environment, even though their return has been so recent (in the timescale of evolution) that they are still making do with bodies that were really designed for land and air!

Fish

Fish are very familiar animals. Too familiar, maybe – most of us carry around a cartoonist's idea of fish as something smooth, silver and slippery – like a trout or a sea-water herring.

But in their own way, each different species is as well suited to its own methods of collecting food as any other form of water life, and the appearance of the fish reflects it.

The stone loach, *above*, is an example. It is quite common in fast streams. It is active at night, spending the day under a stone. It will dart away quickly if the stone is disturbed.

It is a small fish, only about 12 cm long, although it may take five years to grow to that size.

And the grotesque-looking whiskery barbels are used to disturb and grub out freshwater shrimps, worms and insects from the bottom of the stream.

The grotesque **stone loach** opposite *(see 3× life size) is reasonably common, as is the* **bullhead** below *(life size). But neither is easily seen – both of them hide in daytime, and are active and feed at night.*

Look for minnows – small, 'conventional' fish, darkish in colour. They come together in large shoals at spawning time, and the shoals attract other, larger, predatory fish to the spot.

Minnows can easily be trapped in a jam-jar or glass bottle baited with pieces of bread. In fact, that is how most of us humans start our researches into water life!

Look for bullheads – one is shown *below*. They are small fish, about 10 cm long. They prefer clean streams running over gravel. They are dormant during the day, resting like the loach under stones or in among plant stems.

And they are well camouflaged, as you can see, as a defence against predatory birds.

Whether they have started the passion by catching minnows in a bottle or not, it is a fact that more people in Britain *fish* than ever *play* football or cricket or tennis or any other sport. Fishing is the most popular sport. And as you may guess, that has had its effect on the ponds, streams and rivers of Britain.

For a start, many ponds have been preserved (and protected) just for fishing. They are stocked with *coarse* fish such as chubb and roach, which can put up with and breed in turbid water. Trout (which is a *game* fish) can sometimes survive in such water, but will only breed in clean running water. The prime game fish is salmon, which spends adult time at sea, but returns to spawn far upstream in some British rivers.

Birds

Swallows and swifts will often patrol the surface of a pond or river, to catch emerging duns, spinners and other insects.

However, many birds are adapted for a close association with water. The little grebe, *above*, is not as rare as is often thought, although it is a shy bird. (It is commoner in West, Central and Northern Britain.) It is an expert diver, catching small fish, freshwater shrimps, and insects underwater.

It makes a nest from stalks and stems, building a platform, as you see.

Ducks – easily recognisable – are all associated with water. There are many different species, each with its own plumage and way of life, but you can recognise two main groups:

The *dabblers* take their food while at the surface – they up-end to reach down.

The *divers* disappear underwater, to emerge with a plop some distance from where they went down.

The **little grebe** opposite, *the* **coot** below right *and the* **dipper** bottom *are just a few of the water birds to look for. Many of them colonise sewage farms, reservoirs and even waterlogged waste ground close to towns.*

Look for coots – they dive down to bring up a mess of plants which they then pick through. They will also take fish if they can.

They are common birds, related to moorhens, but easily recognisable by their white nose flash, *right*. At nesting time they are aggressive; when two rival males meet, they will each sit back on their tails in the water, and throw up a spray – a kind of mock battle which determines which of the two is stronger (few mating fights cause damage!).

Look for moorhens, identified by a red flash to their beak. They are more likely to be found alone than a coot – and they spend much time on land, but you will often see them walking on a carpet of water-lily leaves if those plants are growing.

Look out for the innumerable ducks that are attracted by open water, even in a town.

And look for swans!

Look for a dipper, *below*; it is an interesting bird. It is quite common. It is a relative of the wren, though somewhat bigger. It is strongly territorial – a pair will patrol a mile of stream. It walks or plunges underwater to feed, grasping the bottom while it collects shrimps and insects. It 'chinks' when disturbed.

Rarities

Kingfishers are not that uncommon, and though not many people have seen one, everybody knows what they look like. Everybody knows what an otter looks like, but unfortunately the chances of ever seeing one in the wild *are* small. For some reason, or probably a combination of reasons, otter numbers have decreased in the last few tens of years. Disease and hunting (although that is now banned) may be partly to blame. But just as important a cause is loss of habitat. Otters like undisturbed banks for their holts, and like many wild animals, they do not breed successfully if they are disturbed.

Nowadays, many streams and rivers are dredged, and their banks straightened and cleared, to improve drainage from the fields and to prevent flooding in winter. Otters remain in dwindling numbers in only a few counties in England, and in Wales and Scotland.

The **otter** opposite *is in a way a signpost to the fact that we are not protecting our wildlife; it is becoming quite rare.* Below, *a modern arrival at the waterside: the* **coypu**, *now naturalised in the Norfolk Broads – it grows to 2 ft (60 cm) long.*

Look for otters. They are rare, but naturalists are trying to find out how many otters we still have. Nobody is quite certain, and any sighting will be very useful. If you do think you have seen, or have evidence of, a wild otter, inform WATCH or your local county Trust for Nature Conservation (find the address from your local library).

Look for otter spraints. Otters mark out their hunting territory – as do many mammals (including the family cat). But whereas the cat uses smelly urine for the marker, the otter uses a 'spraint' – a musky-smelling dropping.

The otter leaves its spraints on exposed places – such as boulders, or on the V-shaped patch of bare ground that you often find where two streams meet. Sometimes it will leave one under a bridge.

The spraint is a splot or globule, dark and gelatinous, tarry looking and it may be shiny with fish bones.

The smell is not unpleasant – it is somewhat sweet. If your 'spraint' does smell foul, it is likely to be from a *mink.*

Mink have escaped from mink farms (where they are bred for their fur) and live wild in the countryside. They sometimes hunt along rivers and can be mistaken for otters.

Look for coypu – another strange inhabitant of our wetlands in East Anglia. It too was brought here (from America) to be bred for fur; and escaped, to live wild. It is large – weighing about 10 kg – and eats sedges, reeds and other vegetation in the Broads, doing much damage to the thick beds. It will also feed on cabbages, sugar beet and other crops in the fields near by.

Adding it all up

'Water is the eye of the countryside.' And to be sure, water is everywhere you look. However, *how* you look is important, and we hope that this **WATCH Book** has given you some new ideas.

Is water a reflection of the land? There is evidence, as we have seen, that we would be wrong if we thought it altogether a completely separate kind of place. Many plants and animals have obviously colonised our fresh water from the land, and keep many terrestrial habits.

But this book has also shown you how to carry out a clean water test (pages 18 and 19); and how to help with some real-life reporting, essential for conservation (page 31). And it has, when you think about it, given plenty of clues on how to stock a pond to keep it thriving: plenty of plants of all kinds (pages 9 to 11), mussels to filter the water and keep it clean (page 20) and lots of pond animals – all of which can be caught by dragging a fine net through an already well-established pond. Indeed, if you are making a pond of your own, 'feed' it with a jam-jar of water from a nearby older pond – the water will be full of mites and other small animals.

So what other hints do you need?